YOUR KNOWLEDGE HAS VALUE

- We will publish your bachelor's and master's thesis, essays and papers

- Your own eBook and book -
  sold worldwide in all relevant shops

- Earn money with each sale

Upload your text at www.GRIN.com
and publish for free

Friederike Lange

Epigenetics in the post genomic era: Can behaviour change our genes?

Was Lamarck just a little bit right?

GRIN Verlag

Bibliografische Information der Deutschen Nationalbibliothek:

Die Deutsche Bibliothek verzeichnet diese Publikation in der Deutschen National-
bibliografie; detaillierte bibliografische Daten sind im Internet über http://dnb.d-
nb.de/ abrufbar.

Imprint:

Copyright © 2011 GRIN Verlag GmbH
Druck und Bindung: Books on Demand GmbH, Norderstedt Germany
ISBN: 978-3-656-01473-7

This book at GRIN:

http://www.grin.com/en/e-book/179170/epigenetics-in-the-post-genomic-era-can-
behaviour-change-our-genes

GRIN - Your knowledge has value

Der GRIN Verlag publiziert seit 1998 wissenschaftliche Arbeiten von Studenten, Hochschullehrern und anderen Akademikern als eBook und gedrucktes Buch. Die Verlagswebsite www.grin.com ist die ideale Plattform zur Veröffentlichung von Hausarbeiten, Abschlussarbeiten, wissenschaftlichen Aufsätzen, Dissertationen und Fachbüchern.

Visit us on the internet:

http://www.grin.com/

http://www.facebook.com/grincom

http://www.twitter.com/grin_com

Embedded in the course Genetics-3H

Epigenetics in the post genomic era: Can behaviour change our genes?

Evolution essay 2011

25/03/2011

Written by Friederike Lange
University of Glasgow
Words: 3,135

Traditionally, genomic research was focused on the investigation of DNA sequence which gives rise to the diversity of phenotypes found in nature. It was undoubted that the information which is given by the genomic sequence is the sole factor which is important for the outcome for each individual organism. But since a few decades, a new concept called epigenetics has arisen and shows that we have to modify our knowledge about genetics. "Epi" derives from Greek meaning "on" or "over" and implies that epigenetic mechanisms act on genes via altering the gene expression and regulation without modifying the actual DNA sequence. In epigenetics, we can find the reason why twins (who have the exact same gene sequence) can alter in their phenotype especially concerning their susceptibility to diseases (Fraga et al. 2005; Wong et al.; 2005).

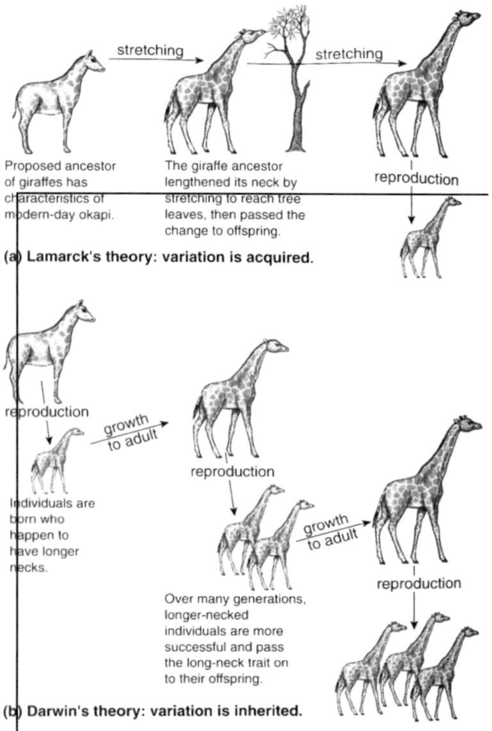

(a) Lamarck's theory: variation is acquired.

Proposed ancestor of giraffes has characteristics of modern-day okapi.

The giraffe ancestor lengthened its neck by stretching to reach tree leaves, then passed the change to offspring.

stretching

stretching

reproduction

(b) Darwin's theory: variation is inherited.

reproduction

growth to adult

Individuals are born who happen to have longer necks.

reproduction

growth to adult

Over many generations, longer-necked individuals are more successful and pass the long-neck trait on to their offspring.

reproduction

Fig 1: Lamarck's (a) and Darwin's (b) theory how giraffes evolved a long neck (Raven and Johnson, 2002)

Furthermore, it has the potential to answer the question how phenotypic characteristics can alter between generations without a change in the underlying genetic material.

Lamarck vs. Darwin:

Broadly seen, the idea behind epigenetics is not new. It appeared 200 years ago, established by Jean - Baptiste **Lamarck** (1744-1829). He was a French naturalist and proposed the theory of inheritance of acquired traits, i.e. that evolution occurs when the characteristics from parents, which they have acquired during their lifetime (induced by environmental changes which in turn changed their behaviour), pass on to their offspring. He thought in particular of the use (or disuse) of specific organs (like the giraffe's neck, see Fig 1) that would lead to its gradual functional improvement (or disappearance). He postulates that these improvements are inheritable and would pass on through generations (so that e.g. the giraffe's neck becomes longer and longer, Fig 1). He believed that evolution is mainly driven by non-randomly acquired, beneficial phenotypic changes which, he said, are inheritable (in particular, those directly affected by the use of organs). However, his theory was discredited to his time and had been displaced by Darwin's theory of natural selection in the early 20th century. **Darwin** (1809 – 1882) ascribed a greater importance to random, undirected changes providing material for natural selection. Both had in common that inheritance of acquired characteristics played an important evolutional role. But Lamarck thought that an organism's inner need drives evolution, whereas Darwin believed devoutly that natural selection of genetic alterations drives the change in evolution which leads to adaptive characteristics in organisms. Thus, Lamarck's theory became decried. But by the recent increase of epigenetics' popularity, his name has come back into modern science. Epigenetics is the field which is now

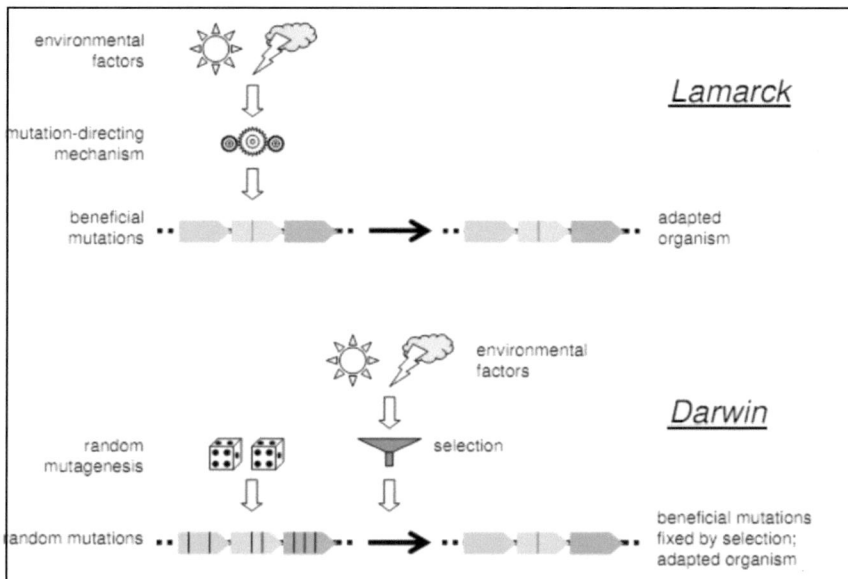

Fig. 2: Lamarckian and Darwinian modalities of evolution. (Koonin and Wolf; 2009)

suggesting that epigenetic changes, rather than genomic changes in the DNA, play an adaptive role for each organism by heritable transmission of acquired characteristics; although, it does not support Lamarck's overall concept. It would be carried too far to say that Lamarck's concept of elongated giraffe's necks is true. But it is put that environmental factors, such as temperature, can influence epigenetic marks. If one puts Lamarck's concept into modern genetics, it would say that "1) environmental factors cause genomic (heritable) changes; 2) the induced changes (mutations) are targeted to a specific gene(s) and 3) the induced changes provide adaptation to the original causative factor" (Koonin and Wolf; 2009). But it has nothing to do with the inner need of the organism as Lamarck supposed. The specific environmental factor must cause an adaptive reaction by a molecular mechanism that triggers the genomic change. And this, in turn, stays in contrast to Darwin, who thought that the environment gives a selective force that

could fix the random changes in the genome (see Fig 2).

Epigenetics: Definition and mechanisms:
Nowadays, there are numerous definitions but "precise definitions have remained both controversial and elusive" (Ho and Burggren; 2010). According to *Berger et al* (2009) "an epigenetic trait is a stably heritable phenotype resulting from changes in a chromosome without alterations in the DNA sequence" These alterations or rather the mechanisms by which epigenetics works are diverse and they are involved in processes like gene regulation, development and even diseases like cancer. So, is it possible that gene expression can be changed by different epigenetic ways, e.g. genes can get silenced typically by methylation of specific nucleotides so that the genes will not get transcribed.

Methylation and demethylation are two of the main types belonging to the epigenetic mechanisms, in which a methyl group is transferred to or removed from a cytosine

3

mostly at CpG-islands which have an important regulative role for gene expression. The methylation mechanism is also important in **genomic imprinting**, another epigenetic mechanism, by which specific alleles get silenced by differential methyl tagging depending on whether the allele is maternally or paternally inherited (e.g. X-chromosome inactivation). This status is partially maintained by differentially methylated regions within or near imprinted genes. Another epigenetic marker is given by the **modification of histones**, which includes acetylation, methylation and phosphorylation. This is especially important in transcriptional regulation since it influences gene expression by changing the chromatin structure, thereby impeding or facilitating gene activation. Besides, small **non-coding RNAs** (including microRNAs) have also been shown to be involved in epigenetic regulation through chromatin/histone modification by directing the cytosine methylation (Costa; 2008). Normally, these modifications are erased during germ cell development and hence, not heritable. But these alterations are stably maintained and thus, inheritable and can have long lasting effects on the gene expression through generations without altering the actual gene sequence (Anway et al.; 2005). In contrast to the genome, the so called "epigenome" is relatively dynamic and is potentially reversible and can be influenced by different environmental factors, especially during the fetal and early postpartal phase. Furthermore, each cell of an organism must not have the same epigenetic alterations: Distinct CpG-islands are differentially methylated in different cell types forming a cell specific methylation pattern. All in all, with the help of these dynamic epigenetic alterations, the genome is able to respond much faster to environmental changes than by changing its actual DNA sequence. Hence, it provides a rapid mechanism to adapt to the environment; e.g. they are able to change the

phenotype within one generation in the agouti mice which show a dramatic heritable change in their fur colour (Morgan et al.; 1999).

Transgenerational epigenetic/maternal effects:

Environmental influences can have long lasting effects on the epigenetic programming, especially when they act prenatal like in the case of the **agouti mice**, which is a inbred strain and thus, genetically identical. This example illustrates visually the potential effects of epigenetic changes by mainly maternal origins. The agouti mice possess the agouti locus which affects the fur colour. This locus is influenced by the methylation degree of the regulatory DNA region upstream from it. Depending on the number of attached methyl groups, the **normal fur colours** yellow to brown and a mottled combination between both can occur in the agouti mice (see Fig 3).

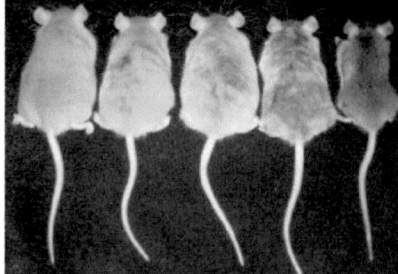

Fig. 3: Five coat colours which represent the variety of differentially coloured agouti mice from yellow, over mottled to brown (Dolinoy et al.; 2006).

If we would think on a strictly DNA sequence dependant heritance of the coat colour, genetically identical parents should give birth to identically looking offspring. But it was observed that those parents (even so they are genetically identical) have different epigenetic states, and they tend to produce differently coloured offspring. Besides, it could have been observed that the progeny of yellow coloured mothers was also mainly yellow coated (respectively: mottled mothers had more mottled offspring) and that these coat colours were mainly influenced by the colour of the mother but not of the father (Waterland and Jirtle; 2003). Therefore, the coat colour is mainly dependant on how much of the methylation marks of the mother have been transferred to the offspring. Furthermore, it was found out that the diet of the mothers can dramatically alter the heritable phenotypic change by changing the DNA methylation pattern and thus, altering the colour of the offspring (Waterland and Jirtle; 2003, Dolinoy et al., 2006). *Waterland and Jirtle (2003)* fed folic acid and other methyl-rich supplements to pregnant mice. The fed methyl-groups were found to bind at the 5' end of the agouti locus, which led to the inactivation of the expression of the agouti gene in the mother and in its progeny. Thus, the offspring of the mice, which received the supplements, had mostly brown coloured offspring, whereas mice without supplements produced mostly yellow offspring, who additionally had an increased susceptibility, e.g. to obesity, diabetes, and cancer. The higher susceptibility for obesity was particular shown by *Dolinoy et al (2006)*, who fed genistein, an isoflavone, as food supplement to pregnant mice which affected gene expression and increased the susceptibility to obesity in adulthood by permanently altering the epigenome. These example shows that an environmental factor in form of nutritional supplement can have a dramatic impact on inheritance and that, without changing the DNA sequence. These effects from the early maternal care taking performances are still detectable in the adulthood of the offspring and can maintain trough life.

If this is also true in humans is not known exactly, but it could explain why twins can have different phenotypes (especially according to diseases) in different environments. In fact, studies with nonhuman primates showed that, when they were exposed to stress in their early life, they developed behavioural patterns which are similar to adolescents who are prone to alcoholism (Heinz et al, 1998). In fact, evidences for epigenetic procedures in humans were found as well: Studies in women during famines show an epigenetic correlation between suffering from starvation *in utero* and the birthweight of their offspring. *Lumey* (1992) could show that mothers, whose mothers suffered from the war induced famine from 1944 – 1945 while being pregnant with them, had offspring with birthweights lower than mothers not exposed to famine. This example shows that even in humans an environmental factor (like famines) can have epigenetic effects on the next generation and, how *Lumey (1992)* showed as well, can even affect the grandchildren. In another study transgenerational epigenetic effects of food supply in the grandparents' prepubertal age were correlated with diabetes and heart disease (Kaati, 2002). They have in common that the most critical time for environmental factors which can induce epigenetic changes and thus, phenotypic changes, is during the prenatal phase.

Behaviour influences and is influenced by epigenetics:

Recently, also evidences for a postnatal influence on the epigenome emerged. But here, the epigenetic changes are not due to chemical exposition (as described before) but rather caused by specific behaviours from the mother which in turn can lead to behavioural changes in the offspring by affecting the chemistry of their DNA. This leads to the hypothesis that the social environment also influences the epigenomic outcome, thereby acting via the same mechanisms as the chemical agents do. The best example is given by the glucocorticoid receptor gene (*GR*-gene) in the hippocampus of rats which is influenced by the maternal care-taking behaviour that can alter the stress response and immune capacity in the offspring: The progeny of rats which showed a higher-than-average care (e.g. measured as rate of licking and grooming) compared to rats showing a care which was below average (showed low rate of licking and grooming) displayed an increased hippocampal expression of *GR*, as well as a weakened hypothalamic-pituitary-adrenal responses to stress (Liu et al, 1997; Meaney and Szyf, 2005; Weaver et al., 2004). It was shown by these researchers that the offspring alter in their DNA methylation patterns (less nurturing leads to less methylation of *GR* and respectively for much nurturing) and that these differences emerge over the first week of life. They also swapped the progeny from high caring mothers low caring mothers, which approved a direct impact of the maternal care (rather than by the DNA sequence) since the effects were reversed. Furthermore, they showed with the help of pharmaceutical that the effects from maternal care on the chromatin structure were reversible (Weaver et al., 2005). Those effects were mainly associated with an altered histone acetylation and transcription factor binding to the *GR*-gene, which were long lasting and persisted into adulthood.

Taking all into consideration, it can generally be said that social behaviour can change the epigenetic programming in other individuals and that these changes can have long lasting effects on the individual behaviour. Therefore, there must be a molecular mechanism that functionally connects maternal care (social behaviour) and the epigenome of the offspring (chemical modification of chromatin and DNA). It is thinkable, that maternal care could induce a signalling cascade in the offspring which in turn activates specific transcription factors leading enzymes to specific target sequences for epigenetically modifying DNA and chromatin. For instance it was shown, that maternal care after birth leads to thyroid hormone-dependant increase of the 5-hypothalamic (HT)-activity at the 5-HT-7-receptor and the subsequently activation

of cAMP and cAMP-dependant protein kinase A (Meaney et al., 2000). It gives a hint how behaviour can lead to changes of DNA-methylations of specific genes. Such cascades must also be inducible by other environmental influences and agents altering one step or more in those pathways which should lead to epigenetic changes (which in turn can lead to behavioural changes as well).

Conclusion:

The idea of a dynamic epigenome is based on the fact that differences between gene function and phenotype are not exclusively determined by the genomic sequence itself, but as well by reversible, but nonetheless, steady epigenetic marks. It gives the ability to the genome to respond quickly to changing environmental impacts by being dynamically regulated. Thus, it provides a more precise and stable control of gene expression and genomic regulation. Single environmental exposures can lead to long lasting effects in the phenotype. Thereby, it is important not to look on the epigenome alone (same is true for the genome) but keeping also the genomic mechanisms (or respectively, the epigenomic mechanisms) in mind, because either will be altered by environmental factors and thereby, will drive evolution by natural selection. A full understanding of the interactions between environment (be it via physical of social impacts) and genes requires that both epigenetic and classical genetic mechanisms are taken into account. It was shown that epigenetic mechanisms and genetic mechanisms are both processes which have long lasting effects on the expression of genes, be it with or without the chaning of the particular sequence. But even the same genetic sequence does not imply that similar epigenetic alterations happen: so was it shown that human twins and other genetically identical organisms do not have the same susceptibility to diseases due to epigenetic differences (especially in the amount of DNA methylation) which occurred over life (Wong et al., 2005). These changes, however, are dynamic and potentially reversible. Moreover, epigenetic changes can be inherited, like it is long known for genetic characteristics.

New evidences show that even behaviour has the potential to interfer with genes by influencing the epigenetic machinerie, especially by maternal determinants (see agouti mice). The possibility that acquired marks can be passed from parents to children has a lamarckain flavour. The idea behind Lamarck is now proved, but in contrast to Lamarck, not the inner need of an animal is decisive, it is more the environmental influence which has the crucial part. But how these interactions between environment and genes are exactly achieved is a purpose of future studies. Main focus should be set on sequence specific and mediating factors that lead enzymes to genes which have the ability to modify histones and DNA and hence, determine the epigenetic status. There must be specific pathways since physiological, pharmacological and behavioural influences can affect these signalling cascades by activation or blocking of specific elements, like changing the chromatin structure and thereby, impairing its methylation and demythelation levels and thus, induce phenotypic changes.

References

Anway MD, Cupp AS, Uzumcu M, Skinner MK 2005. **Epigenetic transgenerational actions od endocrine disruptors and male fertility**. Science. Vol:308. pp. 1466 - 1469.

Berger SL, Kouzarides T, Shiekhattar R, Shilatifard A 2009. **An operational definition of epigenetics**. Genes & Development. Vol:23. pp. 781 - 783.

Costa FF 2008. **Non-coding RNAs, epigenetics and complexity.** Gene. Vol:410(1). pp. 9 - 17.

Dolinoy DC, Weidman JR, Waterland RA, Jirtle RL 2006. **Maternal Genistein Alters Coat Color and Protects Avy Mouse Offspring from Obesity by Modifying the Fetal Epigenome.** Environ Health Perspect. Vol:114(4). pp. 567 - 572.

Fraga MF, Ballestar E, Paz MF 2005. **Epigenetic differences arise during the lifetime**. PNAS. Vol:102(30). pp. 10604 - 10609.

Heinz A, Higley JD, Gorey JG et al 1998. **In vivo association between alcohol intoxication, aggression and serotonin transporter availability in nonhuman primates**. Am J Psychiatry. Vol:155. pp. 1023 - 1028.

Ho DH, Burggren WW 2010. **Epigenetics and transgenerational transfer: a physiological perspective**. The Journal of Experimental Biology. Vol:213. pp. 3-16.

Kaati G 2002. **Cardiovascular and diabetes mortality determined by nutrition during parents' and grandparents' slow growth period**. European Journal of Human Genetics. Vol:10. pp. 682 - 688.

Koonin EV; Wolf YI 2009. **Is evolution Darwinian or/and Lamarckian?** Biology Direct. Vol:4(1). pp. 42 - 56.

Liu D, Diorio J, Tannenbaum B et al 1997. **Maternal care, hippocampal glucocorticoid receptors and hypothalamic-pituitary-adrenal reponses to stress**. Science. Vol:277. pp. 1659 - 1662.

Lumey LH 1992. **Decreased birthweights in infants after material in utero exposure to the Dutch famine of 1944-1945**. Paediatr. Perinat Ep. Vol:6. pp. 240 - 253.

Meaney MJ, Diorio J, Francis D et al 2000. **Postnatal handling increases the expression of cAMP-inducible transcription factors in the rat hippocampus: The effects of thyroid hormones and serotonin.** Journal of Neuroscience. Vol:20(10). pp. 3926 - 3935.

Meaney MJ, Szyf M 2005. **Environmental programming of stress responses through DNA methylation: life at the interface between a dynamic environment and a fixed genome**. Dialogues Clin Neurosci. Vol:7(2). pp. 103 - 123.

Morgan HD, Sutherland HGE, Martin DIK, Whitelaw E 1999. **Epigenetic inheritance at the agouti locus in the mouse** Nature Genetics. Vol:23. pp. 314 - 318.

Raven PH, Johnson GB 2002. **Biology**. New York: McGraw-Hill, 2002. Vol:6.

Szyf M 2009. **Dynamisches Epigenom als Vermittler zwischen Umwelt und Genom.** Medizinische Genetik 1. Vol:21. pp. 7 - 13.

Waterland RA, Jirtle RA 2003. **Transposable elements: targets for early nutritional effects on epigenetic gene regulation.** Mol. Cell. Biol. Vol:23. pp. 5293 - 5300.

Weaver ICG, Cervoni N, Champagne FA et al 2004. **Epigenetic programming by maternal behaviour.** Nature Neuroscience. Vol:7(8). pp. 847 - 854.

Weaver ICG, Champagne FA, Brown SE et al 2005. **Reversal of maternal programming of stress responses in adult offspring through methyl supplementation: Altering epigenetic marking later in life.** Journal of Neuroscience. Vol:25(47). pp. 11045 - 11054.

Wong AHC, Gottesman II, Petronis A 2005. **Phenotypic differences in genetically identical.** Human Molecular Genetics. Vol:14(1). pp. R11 - R18.